# ICME-13 Topical Surveys

**Series editor**

Gabriele Kaiser, Faculty of Education, University of Hamburg, Hamburg, Germany

More information about this series at http://www.springer.com/series/14352

Paul Drijvers · Lynda Ball
Bärbel Barzel · M. Kathleen Heid
Yiming Cao · Michela Maschietto

# Uses of Technology in Lower Secondary Mathematics Education

A Concise Topical Survey

 Springer Open

Paul Drijvers
Freudenthal Institute
Utrecht University
Utrecht
The Netherlands

Lynda Ball
Melbourne Graduate School of Education
The University of Melbourne
Melbourne, Victoria
Australia

Bärbel Barzel
Universität Duisburg-Essen
Essen
Germany

M. Kathleen Heid
The Pennsylvania State University
University Park, PA
USA

Yiming Cao
School of Mathematical Sciences
Beijing Normal University
Beijing
China

Michela Maschietto
Department of Education and Humanities
University of Modena and Reggio Emilia
Reggio Emilia
Italy

ISSN 2366-5947          ISSN 2366-5955   (electronic)
ICME-13 Topical Surveys
ISBN 978-3-319-33665-7    ISBN 978-3-319-33666-4   (eBook)
DOI 10.1007/978-3-319-33666-4

Library of Congress Control Number: 2016939363

Printed on acid-free paper

This Springer imprint is published by Springer Nature
The registered company is Springer International Publishing AG Switzerland

# Main Topics You Can Find in This ICME-13 Topical Survey

- Topical study of the state of the art of the use of digital technologies in lower secondary mathematics;
- Comprehensive survey of research findings;
- Future directions for the use of digital technologies in lower secondary mathematics;
- International perspectives integrated to provide a view on worldwide developments.

# Contents

# Uses of Technology in Lower Secondary Mathematics Education

## A concise topical survey

## 1 Introduction

Digital technology[1] is omnipresent in society. Revolutionary technological developments change the character of professional environments, and therefore put new demands on workers (Hoyles, Noss, Kent, & Bakker, 2010). Consequently, there are new demands on educational systems in order to prepare students for future professions. Importantly, technology also offers opportunities for teaching and learning (see for example, Clark-Wilson, 2010; Sacristán et al. 2010); exploiting these opportunities requires rethinking educational paradigms and strategies. With the advent of such technology, the question arises as to what the impact on education and teaching practices should be in order to prepare the next generation of students for future careers.

Both in professional practice and in personal life, it is particularly striking how digital technologies such as software-controlled engines, smart phones, tablets, and GPS devices rely on mathematical algorithms that are invisible to the user, but play essential roles "under the hood". Implications of these technology-rich environments have the potential to influence the nature of mathematics education and the concepts and skills that future students will possess.

Roberts, Leung, and Lin (2013) comment on the complexity of the interplay between technology, mathematics, and education, noting that this complexity related to the use of tools in mathematics is not a phenomenon that is due to current technologies, but one that has been evident whenever people use tools in mathematics. The rapid development of digital technologies features new capabilities not even considered possible in the past. Despite advances in digital technologies, there is still strong value in using a combination of physical tools and digital technologies in mathematics education (Maschietto & Trouche, 2010). Different types of technologies are available for teaching mathematics, and different technologies are appropriate for different purposes. General technologies for communication,

---

[1]To avoid constantly repeating the terms "digital technology" in this text we will often refer to "technology"; while doing so, we refer to digital technology in mathematics education.

© The Author(s) 2016
P. Drijvers et al., *Uses of Technology in Lower Secondary Mathematics Education*, ICME-13 Topical Surveys,
DOI 10.1007/978-3-319-33666-4_1

documentation, and presentation are essential in order to support the exchange of mathematical ideas. Mathematical technologies, such as spreadsheets, Computer Algebra Systems (CAS), Dynamic Geometry Software (DGS), and applets, enable teachers and students to investigate mathematical objects and connections using different mathematical representations, and to solve mathematical problems (Zbiek, Heid, Blume, & Dick, 2007).

In the context of lower secondary mathematics with current technologies, mathematical procedures have the potential to be outsourced to powerful mathematical technologies (such as graphing tools, spreadsheet software, statistical packages, and Computer Algebra Systems) challenging current curricula goals and teaching and assessment practices. The question arises whether or not the potential change to goals and practices come to fruition in the real classroom. In a study with seven classes of Year 8–10 students, over a period of three years, Fuglestad (2009) found that most students developed the digital competence to make good choices about the mathematical technology (such as different technologies like DGS, spreadsheet, and function plotter) to suit their preferred approaches when solving a given task. Zehavi and Mann (1999) reported an early trial of the use of DERIVE with 13–14-year-old students in which the use of technology changed the organization of the classroom. Technology fostered a change from teacher demonstration followed by student practice of problems, to student control of the modelling process and class discussion following students' work on teacher-prepared computerized tutorials. In a review of studies on integration of Computer Algebra Systems in schools, Barzel (2012) also found that the role of teachers and students changed in the presence of technology.

This topical survey establishes an overview of the current state of the art in technology use in mathematics education, including both practice-oriented experiences and research-based evidence, as seen from an international perspective. We now discuss three core themes related to technology in mathematics education: Evidence for effect; Digital assessment; Communication and collaboration.

- Theme 1: Evidence for effect
  What are the research findings about the benefits of the integration of digital tools in lower secondary mathematics education for student learning?
- Theme 2: Digital assessment
  What are the features of effective digital assessment in the context of both summative and formative assessment and in the delivery of feedback?
- Theme 3: Communication and collaboration
  How can technology be used to promote communication and collaborative work between students, between teachers, and between students and teachers? What are the potential professional development needs of teachers integrating digital technologies into their teaching, and how can technology act as a vehicle for such professional development activities?

In the final section of the survey we offer suggestions for future trends in technology-rich mathematics education and provide a research agenda in light of these trends. We predict what lower secondary mathematics education might look like in 2025, with respect to the place of digital tools in curricula, teaching, and learning, and express some ideas to promote a deep understanding of mathematics.

The issues and findings presented in this topical survey provide an overview of current research, looking forward to a position of effective integration of technology to support mathematics teaching and learning in lower secondary.

## 2  Survey on the State of the Art

In this section we present the state of the art on the use of technology for teaching and learning mathematics in lower secondary education, as well as the research in this field. First, we summarize the quantitative and qualitative evidence of effect (Sect. 2.1) and then digital assessment is addressed (Sect. 2.2). Finally, we discuss the options for communication, collaboration and teachers' professional development (Sect. 2.3).

## 2.1  Evidence for Effect

Over the past decades, there has been considerable research on the impact of technology in learning and teaching mathematics (Blume & Heid, 2008; Clements, Bishop, Keitel, Kilpatrick, & Leung, 2013; Drijvers, Barzel, Maschietto, & Trouche, 2006; Heid & Blume, 2008; Hoyles & Lagrange, 2010). For many teachers, educators, and researchers, new advances in technology and increased access to technology in mathematics education provided opportunities for new perspectives on the development of students' understanding. Many claims have been made concerning the potential for change in mathematics education as a result of the availability of technology and the subsequent benefits for student learning outcomes.

One may wonder, however, whether the potential changes and improved learning outcomes have been realized. If research findings suggest benefits for students´ learning through the integration of technology in mathematics education, this provides an imperative to address the limited use of technologies in lower secondary mathematics courses reported by PISA 2012 (OECD, 2015). Limited use of technology may be due to practical considerations, but it also raises the question of whether the research findings are convincing enough with respect to the benefits for student learning of the integration of digital tools in lower secondary mathematics education. To investigate *whether* digital technology improves student learning, we first revisit some review studies, focusing mainly on quantitative research (Sect. 2.1.1). Next, we address the role of the teacher as an important factor (Sect. 2.1.2). To address the question of *why* technology might improve student learning, this section closes with important results from qualitative research (Sect. 2.1.3).

### 2.1.1 Evidence for Effect: Does Technology Improve Student Outcomes?

The question of the benefits of integrating digital tools in mathematics education, of course, is not a new research question. In the late nineties Heid (1997) provided an overview of principles and issues of the integration of technology and sketched the landscape of the different types of tools and their pedagogical potential. Burrill et al. (2002) reported on 43 studies on the use of handheld graphing technology and concluded that these devices can be an important factor in helping students to develop a better understanding of mathematical concepts; as many of the studies included used qualitative methods, the overall conclusion is not expressed in terms of effect sizes. Ellington (2003, 2006) also focused on graphing calculators. Her review of 54 studies showed an improvement of students' operational and problem-solving skills when calculators were an integral part of testing and instruction, but with small effect sizes. Lagrange, Artigue, Laborde, and Trouche (2003) developed a multi-dimensional framework to review a corpus of 662 research studies on the use of technology in mathematics education and to investigate the evolution of research in the field, but did not explicitly address student learning outcomes. Kulik (2003) explicitly addressed learning outcomes and reported an average effect size of $d = 0.38$ in 16 studies on the effectiveness of integrated learning systems in mathematics.[2] Two subsequent large-scale experimental studies by Dynarski et al. (2007) and Campuzano, Dynarski, Agodini, and Rall (2009), however, concluded that the effects of the use of digital tools in grade 9 algebra courses were not significantly different from zero. Specifically regarding the use of computer algebra systems (CAS), Tokpah (2008) found significant positive effects with an average of $d = 0.38$ over 102 effect sizes. Overall, these studies provide mixed findings on the effect of using digital tools in mathematics education and show different degrees of quantitative evidence.

Let us now focus on three recent review studies that provided information on the effect of using technology in mathematics education through reporting effect sizes, described in Drijvers (2016). The first study by Li and Ma (2010) reviewed 46 studies on using computer technology in mathematics education in K-12 class-rooms, reporting, in total, 85 effect sizes. The researchers found a statistically significant positive effect with a weighted average effect size of $d = 0.28$. This "weighted average" means that it takes into account the number of students involved in each of the studies. Higher effect sizes were found in primary education compared to secondary education and in special education compared to general education. Effect sizes were bigger in studies that used a constructivist approach to teaching and in studies that used non-standardized tests.

The second review study by Rakes, Valentine, McGatha, and Ronau (2010) focused on algebra in particular and reported 109 effect sizes. The interventions were

---

[2]This means that the average difference between experiment group and control group equals 0.38 of their pooled standard deviation, which can be considered a weak to medium effect.

categorized and here we report on two categories: Technology tools and Technology curricula. The average weighted effect sizes for these two categories were $d = 0.151$ and $d = 0.165$, respectively. Over all categories, the authors concluded that interventions that concentrated on conceptual understanding provided about twice as high effect sizes as the interventions focused on procedural understanding. This suggested that the potential of technology was higher for achieving conceptual goals than it was for procedural goals. It was noted that interventions over a small period of time may already have significant positive effect, and also that the grain size differences in interventions (whole-school study versus single-teacher interventions) did not make a significant difference to student achievement.

The third review study by Cheung and Slavin (2013) included 74 effect sizes in K-12 mathematics studies with an average of $d = 0.16$. The authors' final conclusion refers to a modest difference: "Educational technology is making a modest difference in learning of mathematics. It is a help, but not a breakthrough." (Cheung & Slavin, 2013, p. 102). They also found that despite the expected gains due to the development of sophisticated tools, improvement of ICT infrastructure and growing pedagogical experience, the overall effectiveness of educational technology did not improve over time. Similar to Li and Ma (2010), the authors found higher effect sizes in primary education rather than in secondary education. Lower effect sizes were found in randomized experiments compared to quasi-experimental studies. Finally, effect sizes in studies with a large number of students were smaller than in small-scale studies.

Table 1 summarizes the findings of the three review studies. The overall image is that the use of technology in mathematics education can have a significant positive effect, but with a small effect size. Given that any innovative educational intervention usually has a positive effect anyway (Higgins, Xiao, & Katsipataki, 2012), these studies do not provide overwhelming evidence for the effectiveness of the use of digital tools in mathematics education.

The results reported above are mixed, and interpretations reported by authors seem ambiguous. The results of the OECD study show negative correlations between mathematics performance and computer use in mathematics lessons and lead to the conclusion that there is little evidence for a positive effect on student achievement:

**Table 1** Effect sizes reported in three review studies

| Study | Number of effect sizes | Average effect size | Global conclusion |
| --- | --- | --- | --- |
| Li and Ma (2010) | 85 | 0.28 (weighted) | Statistically significant positive effects |
| Rakes et al. (2010) | 109 | 0.151–0.165 | Positive, statistically significant results |
| Cheung and Slavin (2013) | 74 | 0.16 | A positive, though modest effect |

> Despite considerable investments in computers, internet connections and software for
> educational use, there is little solid evidence that greater computer use among students leads
> to better scores in mathematics and reading. (OECD, 2015, p. 145)

A more optimistic conclusion is expressed by Ronau and colleagues:

> Over the last four decades, research has led to consistent findings that digital technologies
> such as calculators and computer software improve student understanding and do no harm
> to student computational skills. (Ronau et al., 2014, p. 974)

As an overall conclusion from quantitative studies, we find significant and
positive effects, but with small average effect sizes in the order of $d = 0.2$. From the
perspective of experimental studies, the benefit of using technology in mathematics
education does not appear to be very strong.

The aforementioned conclusions have some important limitations. First, review
studies are based on studies that are older than the review, which is an issue in an
environment where educational technology and access to this technology is
increasing rapidly. The fact that effect sizes so far have not been increasing,
however, seems to counter this limitation. Second, most review studies concern
experimental, quantitative studies, but do not differentiate between educational
level, type of technology used, the way in which the technology is integrated into
the teaching, or other educational factors that may be influential. Therefore, they
provide an interesting overview, but do not give detailed accounts of individual
studies that may report cases where technology has had a great impact. In short, the
review studies provide an overall picture that helps to answer the question of
*whether* student achievement may benefit from the use of technology, but they do
not provide insight in the reason *why* this might be the case. To get more insight
into this why-question, we now focus on one important factor, namely the role of
the teacher.


## 2.1.2   An Important Factor: The Teacher

There is general agreement in the research that the teacher's ability to integrate
digital tools in mathematics teaching is a crucial factor when working in a classroom
where technology is available. A large body of research identifies essential factors
such as mathematical knowledge, pedagogical skills, pedagogical content knowledge, curriculum knowledge, and beliefs for effective teaching (Adler, 2000; Even &
Ball, 2009; Koehler, Mishra, & Yahya, 2007; Remillard, 2005; Roesken, 2011).

The acknowledgement that teachers need specific knowledge and skills to successfully integrate digital resources into their teaching has resulted in the development
of different frameworks and models for describing teaching strategies and for fos-
tering teachers' professional development in using digital technologies in mathematics teaching. For example, based on the notion of instrumental genesis (Artigue,
2002), the model of instrumental orchestration highlights the importance of the
so-called didactical configuration for effective teaching with technology, and of the
mode in which the teacher exploits such a configuration (Drijvers & Trouche, 2008;

Trouche, 2004). Ruthven (2007, 2009) offers a practitioner model for successful use of technology, identifying working environment, resource system, activity format, curriculum script, and time economy as key components.

The Technological-Pedagogical Content Knowledge Framework (TPACK) is defined as the coherent body of knowledge and skills that is required for the implementation of ICT in teaching (Koehler et al., 2007), but is not specific to mathematics education. Applied to the teaching of mathematics, the acronym M-TPACK is used. Although the idea of TPACK has been criticized (e.g., see Graham, 2011), it has been used regularly to frame research on teachers using technology. As an example of a quantitative study with M-TPACK as a framework, we now describe a recent study from China (Guo & Cao, 2015). The Chinese government has encouraged the use of technology in mathematics education for more than 20 years. It is worth mentioning that the content related to ICT use accounts for 6.34 % of Chinese mathematics curriculum standards (Guo & Cao, 2012).

The aim of this study was to identify decisive factors impacting the use technology on student achievement. The study involved 65 junior high school mathematics teachers and nearly 2500 students from three representative school districts in China, and data for two years were used in this research to detect the effects of teachers' information technology use. The study explored the impact of teachers' use of digital tools on students' mathematics achievement through a hierarchical linear model, taking students' achievement in 2012 as the dependent variable and teachers' M-TPACK, IT usage, students' achievement in 2011 and "shadow education" (i.e. personal tutoring or remedial class) time as the independent variable. Students' data were taken from a longitudinal study entitled Middle-school Mathematics and the Institutional Setting of Teaching in China (MIST-China). Three school districts were selected as representative samples separately in northern, north-eastern, and south-western China.

The dependent variables for this study were student scores in algebra or geometry on an achievement test based on curriculum standard (equal interval, item-response-theory scaled) and administered to grade 7 and grade 8 students in the sampled schools in May 2011 and 2012, respectively. Students had 40 min to complete twelve problems in algebra or geometry. Based on the scale in Survey of Preservice Teachers' Knowledge of Teaching and Technology and developments (Landry, Anthony, Swank, & Monseque-Bailey, 2009), an M-TPACK scale was developed and used (Guo & Cao, 2015).

The students' questionnaire included items that measured students' shadow education time and social economic status (SES). Since, for the most part, there was no change in students' SES over a period of one year and the prior scores consisted of effects of SES, SES was not a dependent variable in this study. The teachers' questionnaire included items that measured teachers' background information (such as years of teaching service), what ICT has been used in classroom teaching, and the frequency of the use of ICT.

The results showed that students' prior mathematical ability was important to students' mathematics achievement. With corrections for the effects of students' prior achievement and shadow education, the teachers' TPACK had significant positive effects on the students' mathematics achievement in both algebra and geometry; the effect on geometry was greater than on algebra. This supports the conjecture that teachers' ability to integrate digital tools in mathematics education is indeed an important factor for the benefit of that integration.

### 2.1.3  Evidence for Effect: Reviewing Qualitative Studies

As developed in the previous section, quantitative studies are valuable in that they offer a type of knowledge sought by policy makers–knowledge that offers statistically conditioned conclusions about the effects of particular instructional or learning conditions. This is what we called the *whether*-question. However, learning mathematics using digital mathematical tools raises fundamental questions about the type of reasoning accommodated by the tools and the role of representations in the context of that tool use. These questions are important considerations for students at ages 10–14, a critical age span during which Piaget placed the development of cognitive abilities that begin to crystallize for many. But quantitative studies are not particularly suited for probing deeply into the nature, whys, and wherefores of teaching or learning mathematics.

By focusing on what students do with mathematical technology and probing students' thinking as they engage in mathematics in the presence of mathematical digital tools, qualitative studies afford researchers increased opportunities to understand the nature of the effects of digital tools on mathematics teaching and learning. As important as the results of such qualitative studies are the constructs about teaching and learning that can be developed in the context of those studies; they help us to answer the *why*-question.

As was revealed with the advent of the first affordable personal computers, the integration of digital mathematical tools in school mathematics classrooms has the potential to alter both what mathematics students learn and how they learn it (Fey et al., 1984). Although technology can make the learning of particular mathematical content more easily accessible, it can also make that learning problematic. Drijvers' (2004) extensive study of the learning of algebra in a computer algebra environment accentuated this point. Qualitative studies such as Drijvers' have generated and tested new constructs for learning mathematics using digital tools. While Drijvers' study documented that students in his study increased their understanding of some roles of parameters, by using the construct of instrumentation (i.e., the ways in which students' thinking must accommodate the technology with which they are working) it also uncovered the intricacies of students' experience. For example, seemingly straightforward procedures to solve equations with computer algebra turned out to provide both technical and conceptual obstacles to students, and the two types of obstacles were shown to be related. The rich and thorough nature of the qualitative data that underpinned this study, coupled with the researcher's sensitivity to the relationship between the student and the tool, enabled the study to advance the field's understanding of how the constructs of instrumental genesis and instrumentation schemes apply to computer algebra systems.

The classroom availability of technology such as dynamic geometry software (DGS) and the capacity of these tools to readily generate numerous instances enabling students to conjecture relationships have highlighted the need to examine students' conceptions of the role of evidence in the establishment of mathematical truths. Hadas, Hershkowitz, and Schwartz (2000) designed and conducted a qualitative study with middle school students to investigate the nature of this behavior in the context of technology-based activities intended to cause surprise and uncertainty. The qualitative design of the study required researchers to account for the explanations students gave, rather than merely determining whether students' explanations matched the ones researchers had expected students to offer, resulting in the researchers identifying a previously unexpected genre of explanations (visual/variational) that either were based on the dynamic displays or stemmed from students' (presumably DGS-based) imagery. A possible new norm for mathematical explanations was discovered.

A qualitative approach that also focused on accounting for student explanations led Leung (2015) to highlight the importance of dragging to discover invariant properties of constructions, starting from the assumption that "*variation in different aspects of a phenomenon unveils the invariant structure of the whole phenomenon*" (p. 452). An important feature of DGS is the capacity to visually represent geometrical invariants, as well as simultaneous variations induced by dragging. For geometrical configurations, students may perceive the variations of the moving image through contrast to what simultaneously remains invariant. The author distinguishes two levels of invariants and different drag modes. Discerning invariants and invariant properties can lead a learner to transform acting into perceiving conceptual and theoretical aspects of Euclidean geometry.

As a different way to exploit the dragging option in DGE, Falcade, Laborde and Mariotti (2007) introduced the notion of function in terms of variation and covariation. These meanings are fostered through exploration of the effect of particular Cabri macro-constructions where the movement of a point (P in Fig. 1) causes the movement of another point (H). In particular, the authors used the Trace tool in order to foster the emergence of twofold meaning of trajectory, namely as a globally perceived object and as an ordered sequence of points. The students' answers showed the use of the dragging to identify the nature of variables and, at the same time, the domain and the range of a function as a trajectory.

Qualitative research on the use of digital mathematical tools in the learning and teaching of mathematics has not been limited to the use of a single tool. Kaput's SimCalc research program tied mathematics technology to technology that facilitated communication and connections across users. In a 26-chapter book (Hegedus & Roschelle, 2012), researchers focused on examining the learning of middle grade students as they engaged in networked activities using simulations and multiple represented movements (e.g., races and elevator rides) to develop an understanding of important mathematical underpinnings of calculus. One such qualitative study (Bishop, 2013) focused on student learning and based conclusions on analysis of video footage of the same SimCalc curriculum unit across thirteen seventh grade

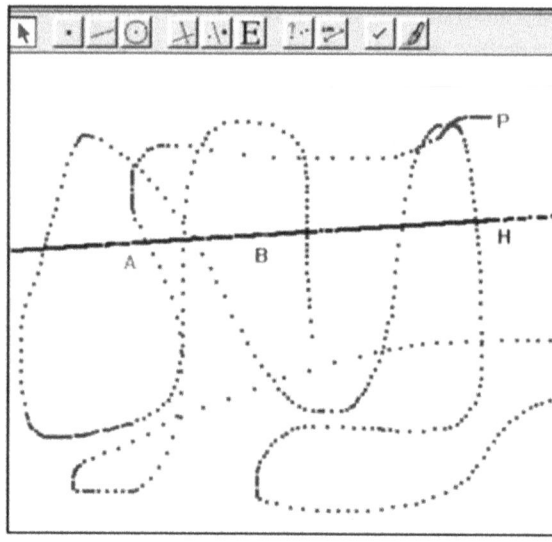

**Fig. 1** Screenshot of the trajectories of points P and H (Falcade et al., 2007, p. 324)

classrooms. This collection of qualitative data allowed the researcher the unusual opportunity to develop a qualitative synthesis of the intellectual work in this use of technology. The researcher concluded from her qualitative synthesis that discourse mediated not only teacher-student interaction with the technology but also with the underlying mathematics.

In other projects, digital technology is used to share problem solving strategies developed in group work and to make or manipulate digital animations of physical objects. We discuss two different situations involving the Pythagorean theorem. In the first situation, Anabousy and Tabach (2016) proposed to three pairs of seventh-grade students a GeoGebra applet and an inquiry task (see Fig. 2): Do you think there are relations between areas of squares built on the sides of an obtuse/acute triangle? For the authors, the interesting element in the analysis is that during their exploration two students discovered the Pythagorean theorem.

From another perspective, Maschietto (2016) presents a teaching experiment in which seventh-grade students approached the Pythagorean theorem first by

Do you think there are
relations between areas of
squares built on the sides of
an obtuse/acute triangle?
Explain!

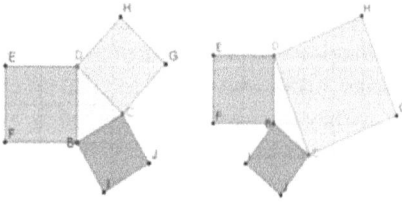

**Fig. 2** The task for the students (Anabousy & Tabach, 2016, p. 2)

**Fig. 3** Wooden model (*left*), students during discussion (*center*), and IWB (*right*) (Maschietto, 2016, pp. 2 and 4)

manipulating a wooden model (Fig. 3, left) in groups, and next by collective discussions in which the model is reconstructed on an interactive whiteboard (IWB). The use of the IWB enables a new collective manipulation of the machine (Fig. 3, center), in which the movement of the wooden model is shared and emphasizes the conservation of areas. It also supported the students' argumentation processes (Fig. 3, right), taking into account students' difficulties.

Qualitative studies have highlighted the subtlety of using digital technology in lower secondary mathematics education, and have inspired the development and refinement of theoretical constructs that explain the opportunities and the pitfalls. With the continuing exponential growth of initiatives for technologizing mathematics classrooms, there is a burgeoning need for the development and refinement of additional theoretical constructs that guide the design of digital-technology-intensive mathematics experiences for students, especially middle school students who are on the verge of developing increasingly sophisticated mathematical thinking, and that inform day-to-day classroom practices.

## 2.2 Digital Assessment

Assessment plays an important role in education, and mathematics education is no exception to this. Student assessment can take a variety of forms and can be either summative or formative (Black & Wiliam, 1998; EACEA/Eurydice, 2011; Wiliam, 2011). In summative assessment, the goal is to evaluate student learning, skill acquisition, and achievement at the end of instruction and often serves as a gateway to successive levels. Formative assessment concerns "the process used by teachers and students to recognize and respond to student learning in order to enhance that learning during the learning" (Bell & Cowie, 2001, p. 540).

Traditionally, assessment in mathematics, and summative assessment in particular, has been constrained to pen-and-paper tasks. All students complete the same tasks at the same time. Student work is usually graded by a human assessor, in many cases the student's mathematics teacher. Grading often makes use of partial credit: if a student makes a mistake in one out of a series of steps or manipulations, the assessor can decide to assign a part of the full credit available for the assignment as a whole. Digital assessment provides new opportunities for summative and formative assessment of mathematics and questions the traditional assessment paradigm. The change of medium from offline to online may have an impact on the nature of tasks, on the grading, and even on the type of abilities and skills assessed. A natural question, therefore, is how digital testing—in many cases delivered and administered online—affects the type of skills assessed, the goals of the assessment, the tasks, and the validity and the reliability of the assessment.

Stacey and Wiliam (2013) distinguish two types of technology-rich assessment, which we call assessment *with* technology and assessment *through* technology. Assessment with technology can be a traditional written test, during which students may use technology such as (graphing or CAS) calculators. Such a model is used in final national examinations in many countries (Brown, 2010; Drijvers, 2009). We speak of assessment through technology when technological means are used to deliver and administer assessment. In many cases, the latter comes down to an online test. The focus in this section is mostly, though not exclusively, on assessment through technology.

Stacey and Wiliam (2013) make explicit the ways in which technology is influencing the assessment: The teacher may outsource selecting and presenting tasks, as technology allows automated generation of similar tasks or may enable new types of tasks such as drag-and-drop items or dynamic situations to be analyzed. These new opportunities provided by technology have the power of questioning the traditional assessment paradigm, as tests can be more easily individualized to address a wide range of competencies to meet individual needs. Also, technology can impact the way students operate and reason when working on tasks, for example while using CAS to solve equations or while exploring a dynamic construction with a geometry package to develop a conjecture that may be proved. Relieving students from computation and drawing affects the type of skills assessed, the goals of the assessment, the tasks, and the validity and the reliability of the assessment. Students' mathematical literacy abilities can be assessed more easily, as well as their conceptual understanding, strategies, and modelling and problem-solving skills (Stacey & Wiliam, 2013).

Technology also influences the way teachers can deal with students' responses as well as their administration of results, as technology can provide detailed information about individuals' strengths and weaknesses, in the form of an overview or an individual diagnosis (e.g., see Fig. 8). This type of information can provide diagnostic feedback about specific competencies and when provided to students diagnostic feedback can foster individual learning.

To address the questions on the affordances and constraints of digital assessment of mathematics, we will discuss some general points including validity, and discuss the case of summative assessment in Sect. 2.2.1. In Sect. 2.2.2, we will elaborate on the formative assessment of mathematics.

### 2.2.1 Digital Summative Assessment

As Stacey and Wiliam (2013) have shown, there are strong arguments for digital assessment. For assessment *through* technology, the following arguments hold, some of which go beyond the topic of mathematics:

- The delivery argument: Compared to a paper and pen test, digital assessment makes it easier to deliver a test at different moments or in different places. Taking the test, therefore, becomes less time and place dependent. Delivery to a large number of students is also relatively easy; scaling up is no longer as complex.
- The production argument: If an extended task database is available, it is relatively easy to produce a new version of a test that is comparable to previous versions. In particular, when large numbers of students are involved and psychometrical item data are gathered, test quality can be controlled in a sophisticated way that would be much harder, if not impossible, to achieve through paper-and-pen media.
- The feedback argument: Digital assessment allows for the automated generation of feedback. Such feedback can be purely technical or focusing on the mathematical content. Hints and other forms of support may be delivered; the student's use of these features may be incorporated in the calculation of the student's score.
- The scoring argument: Grading of students' work may be automatized. This may save much time for the teacher. In addition, automated grading is not only fast, but also objective and consistent; its results may provide data for learning analytics.
- The adaptive argument: During the administration of a digital test, the type of feedback or the type and level of subsequent assignments can be adapted to the student's results so far. If the digital assessment system keeps track of the student's progress through an automated evaluation and an updated student model, the adaptive test can focus efficiently on the appropriate level for the student and hence come to a diagnosis or decision using a less time-consuming test.

For assessment *with* technology, a first argument is that it is common practice nowadays to do mathematics with the support of technology. A written test without access to technology, therefore, would be an anachronism. A second argument is that the use of technology allows for outsourcing laborious calculations, so that assessment time can be used not just to assess basis skills, but also to focus on higher order thinking skills such as conceptual understanding, modelling, and problem solving.

These arguments make a strong case for the assessment of mathematics with or through digital means. In practice, however, the realization of the above features is still cumbersome and the limitations, particularly in the case of assessment through technology, may result in questioning the validity. In line with Wools, Eggen and

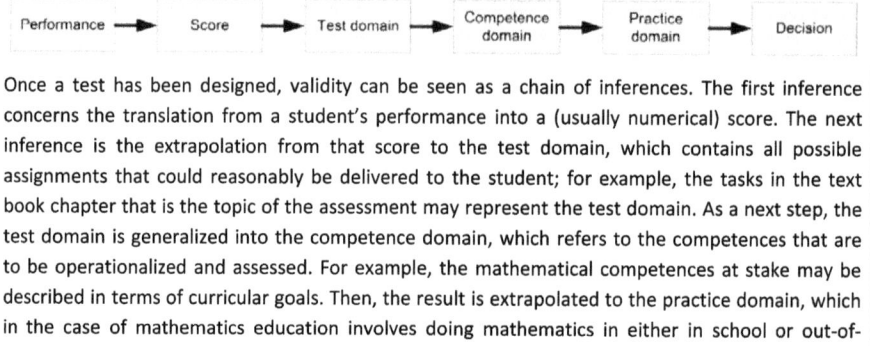

Once a test has been designed, validity can be seen as a chain of inferences. The first inference concerns the translation from a student's performance into a (usually numerical) score. The next inference is the extrapolation from that score to the test domain, which contains all possible assignments that could reasonably be delivered to the student; for example, the tasks in the text book chapter that is the topic of the assessment may represent the test domain. As a next step, the test domain is generalized into the competence domain, which refers to the competences that are to be operationalized and assessed. For example, the mathematical competences at stake may be described in terms of curricular goals. Then, the result is extrapolated to the practice domain, which in the case of mathematics education involves doing mathematics in either in school or out-of-school. Finally, the inference leads to a decision, which can be a pass or a fail, a grade for the student's level of competence, a suggestion on how to proceed the learning process, or a diagnosis of weak and strong points in the learning so far.

**Fig. 4** Validity as a chain of inferences (Wools et al., 2010, p. 65)

Sanders, we see validity as "a chain of inferences that are made to translate a performance on a task into a decision on someone's abilities or competences" (Wools, Eggen, & Sanders, 2010, p. 65). Figure 4 depicts such a chain of inferences and may be helpful to identify possible constraints in digital assessment.

As a first remark, we notice that a good test requires a profound analysis of the epistemological and didactical aspects of the competence at stake. This is always a subtle matter, but in the case of testing through technology an additional aspect is the implementation of the results of this analysis in a student model, that can be implemented in the digital assessment system (Chenevotot-Quentin, Grugeon-Allys, Pilet, & Delozanne, accepted; Grugeon, 2016; Grugeon, Chenevotot, Pilet, & Delozanne, 2013). During the delivery of a test, the student's position in the student model, which may include subtle and high level competencies, should be constantly monitored and updated, in order to allow for adaptive testing. In terms of Fig. 4, this epistemological and didactical analysis should guarantee a sound inference from test domain to competence domain. Even if this is not really a constraint of digital assessment, it is an important point of consideration.

A second possible constraint of digital assessment is that learning goals usually include higher order thinking skills such as problem solving, modelling, reasoning, and proving. Within the constraints of current digital assessment environments (limited construction room for students, hardly any options to interpret reasoning or proof), it is not easy to assess these competences through digital means. In some cases, the test is restricted to basic skills, in comparison to the higher goals of the competence domain, as expressed, for example, in curriculum documents. In terms of the model in Fig. 4, this constraint may affect the inference from test domain to competence domain.

A third constraint concerns scoring. In many cases, human assessors make use of partial credit, for example, when they notice that a minor procedural mistake led to an incorrect answer, but that the overall approach of the problem is worth a partial score. Most digital assessment systems, however, are unable to carry out such a refined step-wise automated scoring for mathematics. Even if automated scoring is becoming more and more sophisticated and if the difference with human scoring is decreasing (VanLehn, 2011), the refined step-wise human scoring that is traditionally carried out in mathematics tests and comes with the use of partial credit, is not yet achieved in assessment through technology. This constraint relates to the inference from test domain to competence domain in Fig. 4.

A fourth constraint concerns the limited room for students during digital assessment to "do mathematics". Doing mathematics, in addition to thinking, often comes down to activities such as scribbling on a piece of paper, sketching a graph, puzzling with formulas, roughly drawing a geometric figure, and schematizing a situation or a problem-solving process. These activities may be easier to carry out on paper rather than with a digital tool, particularly if powerful means such as formula editors, graphing tools, spreadsheets, computer algebra, and dynamic geometry systems are lacking. Consequently, the work carried out within an assessment environment does not reflect authentic mathematical practice, as students cannot do what they are accustomed to doing and have no room for construction or production because of the system's limitations. Of course, paper and pen may be used to overcome these limitations, but this is not practical if the goal is for assessment within a technology and in this case only part of the student work would be visible within the assessment system. In terms of the model in Fig. 4, a sound inference from competence domain to practice domain requires that the competence domain reflects practice, in this case doing mathematics in some sort of in-school or out-of-school reality. The assessment system's constraints with respect to test item types may also be an obstacle for testing mathematics as it is done in practice. One approach to deal with this is to use artificial questions to overcome the system's constraints. Figure 5, for example, shows a test item in which the square

In the triangle below, $\cos(\alpha) = ¼$. $BC = \sqrt{k}$

Enter the value of $k$: _____

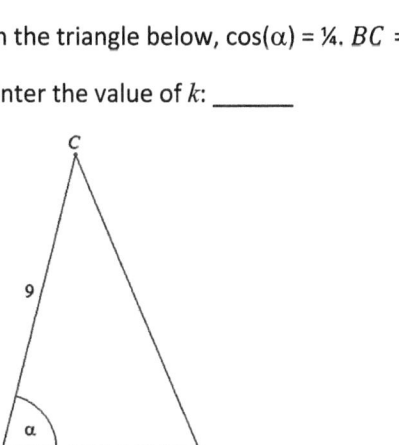

**Fig. 5** Artificial question in a digital test to avoid the assessment system's constraints

root of a length is asked in order to avoid the system's inability to deal with expressions such as $BC = 3\sqrt{10}$.

Some criteria have been suggested for assessment systems that help to overcome constraints (e.g., see Bokhove & Drijvers, 2010). These criteria concern both assessment with and through technology:

1. The user can enter formulas and expressions through a formula editor;
2. The user can graph functions and explore them;
3. The user can make tables of function values;
4. The user can carry out geometrical constructions;
5. The system is connected to an expert system such as a computer algebra system so that student input can be interpreted in an "intelligent" way and that scoring and/or feedback can be delivered in a close-to-human way;
6. The system can deal with a student model and can keep track of the student's positioning in it during the delivery of the test.

Of course, this is a demanding set of criteria and we are happy to say that important achievements have been made recently in these directions. In the meanwhile, we believe that too crucial constraints, as they exist in assessment systems, may challenge the validity of assessment of mathematics through technology and an urgent agenda is to turn these challenges into achievements.

An example that illustrates summative assessment *with* technology is provided in Ball (2015) where a cohort of Year 12 students was allowed to use CAS in their end-of-year examinations. There was a range of problems on the examinations. For some problems use of CAS was not beneficial, for example, because students had to show why a result was true or because the problem involved a routine task where CAS might be expected to take over pen-and-paper work which would be too time consuming. An analysis of Year 12 examination scripts for these students showed that for some problems many students chose to outsource procedural work to CAS, which suggests that students were making judgements about the efficiency of various techniques available. This study highlights the importance of teachers and students discussing choices about use of technology or pen-and-paper in assessment and the need to focus on efficiency of both, depending on the nature of problems.

### 2.2.2 Digital Formative Assessment

The aim of formative assessment is the adaptation of learning and teaching practices to fit students´ needs. Two steps are important for formative assessment: gathering information about the students' achievement and forming appropriate next steps to improve the achievement. Wiliam and Thompson (2008) conceptualized this process in five key teaching strategies:

– Clarifying/understanding/sharing learning intentions and criteria for success;
– Engineering effective classroom discussions and other learning tasks that elicit evidence of student understanding;

– Providing feedback that moves learners forward;
– Activating students as instructional resources for one another;
– Activating students as owners of their own learning.

These strategies are described from the perspective of the teacher engineering a lesson with the specific aim of assessing students' achievement as an important strategy for realizing formative assessment and for highlighting the goal of making the students owners of their own learning. These strategies are also crucial for other modes of formative assessment, such as self- or peer-assessment, that are not whole class assessment.

As an example, the FaSMeD-project (Formative assessment in science and mathematics education, https://research.ncl.ac.uk/fasmed) tries to classify existing ways of formative assessment and comes up with three dimensions to classify different types of formative assessment. One dimension concerns the strategy dimension according to the above list by Wiliam and Thompson (2008); the second dimension concerns the agents as the responsible persons conducting the assessment, and the third dimension describes the role or functionalities of technology which is used (Aldon et al., in press).

Audience response systems (ARS), which allow students to give answers via mobile or offline-handhelds are engineered by the teacher and provide an immediate overview of the students' answers. The technology is used to send tasks or share answers, as well as to collect and analyze the answers. Mostly the answers are not identified with individuals, but instead help the teacher by providing feedback about the class or as an impetus for classroom discussion.

In self-assessment arrangements, the agents are either single students or peers. An intensive approach to self-assessment is shown in the following example by Ruchniewicz (2016), which is programmed using TI-Nspire and has been developed in the frame of FaSMeD. The functionality of technology is mainly to provide tasks and mathematical information. The students become their own assessors according to content that they should have already learned in the classroom. The start is an open task, here the transfer from a situational to a graphical description of a functional relation. The required graph can be drawn by dragging and dropping parts of the graph. Instead of getting a score as immediate feedback, students are provided with a checklist to enable self-assessment of achievement (Fig. 6); one check is, for example, "Does your graph meet the $x$-axis three times?" After every check item, the student has the chance either to get supportive information or further exercises to overcome a specific misconception. Through these opportunities, the students get a chance to re-acquire knowledge themselves and maybe overcome existing gaps.

The benefit of this technology in comparison to a paper-and-pencil version is that the student can focus on what s/he needs for their personal learning trajectory and does not have to manage all the other material. Although it is unusual for students not to get immediate feedback and scoring, they appreciate having the chance to get support to overcome their gaps in knowledge by themselves; this has been observed occurring successfully in many cases. The trajectories of the

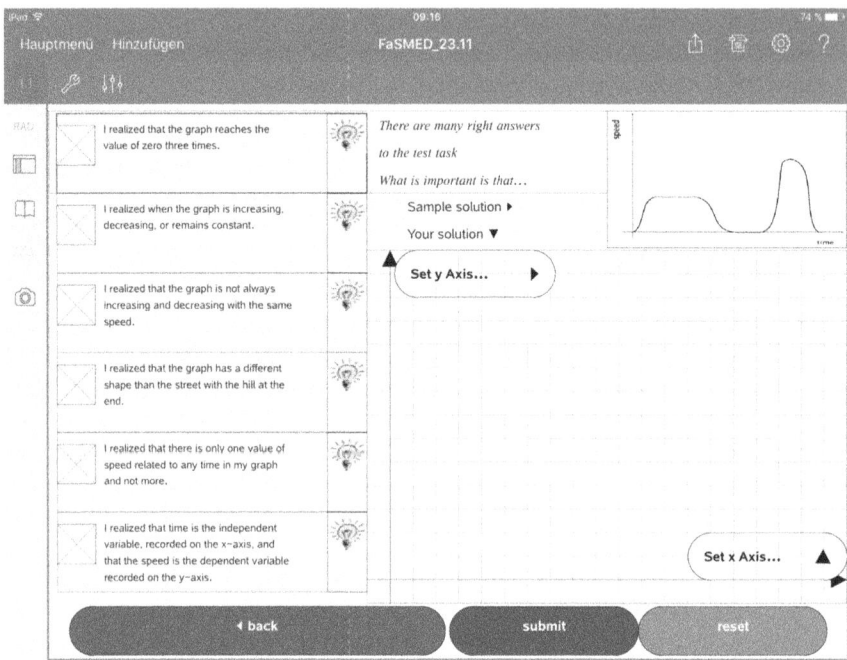

**Fig. 6** Technology providing a checklist for self-assessment from the FaSMeD project

individual students are provided for the teacher in a spreadsheet file to give an impression of the students' achievement.

The following example is taken from a diagnostic test in the Netherlands for high achieving 14–15 year old students (see https://www.hetcvte.nl/item/diagnostische_tussentijdse_toets). The test is done both with and through technology: technology not only delivers it, but also offers means for interactively doing mathematics. Think of the following task: Enter an equation of a parabola that has the point (3, 2) as its vertex. The assessment tool provides the student with an equation editor in which "$y = $" is already pre-set (see Fig. 7).

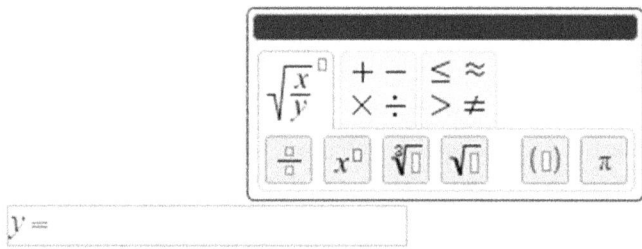

**Fig. 7** Formula window with editor palette above from a Dutch online assessment system

Now the student, assuming that he or she is familiar with formula editing, can complete the equation in many ways, such as $y = (x - 3)^2 + 2$, which is correct. There are, however, other correct answers, such as $y = x^2 - 6x + 11$ or $y = -10(x - 3)^2 + 2$. To guarantee an appropriate diagnosis, algebraic intelligence is needed to distinguish the infinite number of correct responses from the equally numerous incorrect solutions. Interpretation of the student's response should also identify possible mistakes. For example, the response $y = (x + 3)^2 + 2$ suggests an understanding of the notion of the equation of a parabola and the relationship with its vertex, but a weak procedural fluency in the field of positive and negative values while translating. In the case of this test from the Netherlands, this type of intelligence is provided by a computer algebra system running at the back-end of the assessment platform. In addition to providing appropriate tools for the student (in this case the formula editor) and for the interpretation of responses (computer algebra in this case), the system should be able to interpret results in terms of a conceptual student model that allows for reporting a diagnosis. Figure 8 shows an example of such a diagnosis, based on a two-dimensional model consisting of mathematical domains and competences.

This example suggests the following key features for digital diagnostic assessment. First, students need appropriate tools to express their mathematical thinking. In the example, this was a formula editor. Second, intelligence is needed to interpret

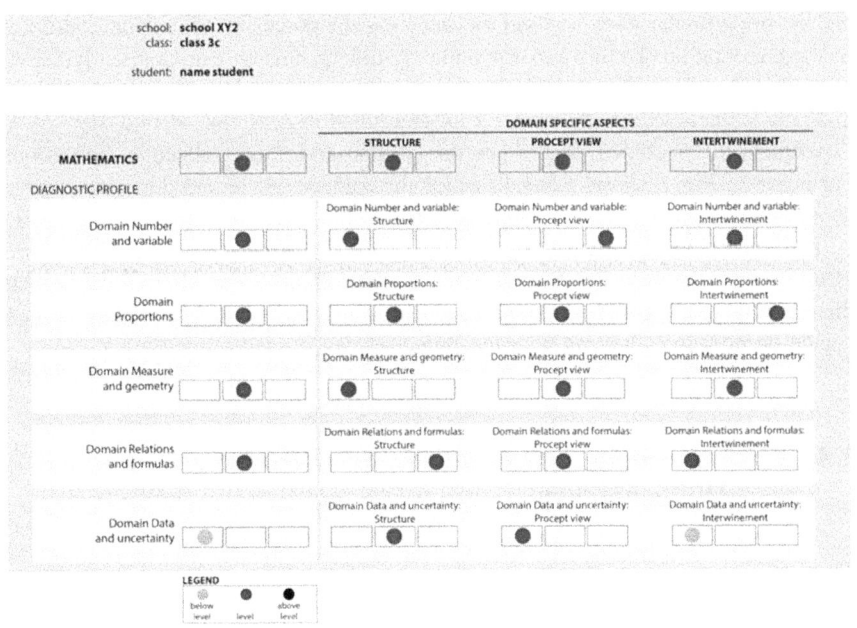

**Fig. 8** Automated diagnosis to the student, adapted from http://www.pilotdtt.nl/over-de-dtt/nederlands-engels-en-wiskunde

In the diagram ABC is an isosceles triangle (AB=AC)

DA, DB and DC are the bisectors of angles A, B and C correspondently

1 Gil conjectures that the bisectors always divide the triangle into three congruent trianlges

Generate a counter example to Gil's conjecture and click ⸢Submit⸣

or click ⸢none⸣ if there is no counter example

2 Shai conjectures that one of the triangles created by the bisectors can be an acute triangle

Generate a supporting example to Shai's conjecture and click ⸢Submit⸣

or click ⸢none⸣ if there is no supporting example

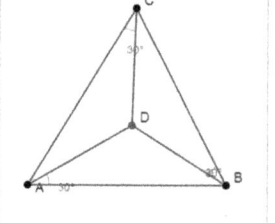

**Fig. 9** Geometry task by Luz and Yerushalmy (2015, p. 151)

student responses for the sake of diagnosis. In the example, computer algebra provides this expertise. In general, step-wise tracing and interpreting of student work is needed, for example with the help of so-called domain reasoners (Heeren & Jeuring, 2014). Furthermore, a student model is needed to frame the diagnosis, as well as a means to produce reports that inform the student (and the parents and the teacher) on his/her achievements and to provide a diagnose of the current level and the points that require extra action or practice. An indication of what causes mistakes, the why behind them, is needed to get a grip on the student's thinking and to open ways for improvement.

In their study, Luz and Yerushalmy (2015) took on the challenge of automated assessment of knowledge of geometric proofs for middle school students (grades 8 and 9). In particular, they focused on assessing the processes of making conjecture and argumentation. To this aim, the authors implemented an examination including tasks that focus on the comprehension of terms and reading the proof. A DGS diagram is included in the task. This examination was the second step of an experiment that included, first, a practice session and, then, collective discussions. For instance, Fig. 9 shows a task in which the students are required to give answers about counter-examples of the statement they were asked to explore using DGS.

Through using a group feedback sheet, this study seems to show the affordance of this assessment tool for informing the teacher about students' interpretations of the elements in the given geometric statement. This may serve as a starting point to decide on how to follow up with teaching.

## 2.3 Communication and Collaboration Through and Of Technology

Technology access can enable a rethinking of teaching and lead to increased pedagogical opportunities (see for example, Pierce & Stacey, 2010). Communication and collaboration are essential aspects for this rethinking (Yackel, 2002; Woo & Reeves, 2006) in order to deepen conceptual and procedural knowledge in students.

Communication can be twofold when learning with technology: communication *through* technology and communication *of* technology (i.e., the technology output). Both types of communication—*through* and *of* technology—can occur simultaneously and they can happen either in the classroom or online between collaborators at different locations. One could consider the entry of syntax or programming of technology as communication *with* technology, as it requires an interaction between a person and a technology. However, we are not classifying this type of interaction as 'communication' in this paper. Our rationale for this decision is that we are using the term 'communication' to represent a social interaction and not an individual learning situation.

Communication *through* technology involves the use of display technologies, both hardware and software, to present and exchange mathematical ideas. These technologies enable the display of student/teacher screens, either statically or dynamically, providing data for classroom discussion (see for example, Clark-Wilson, 2010; Guin & Trouche, 1999). We include presentation programs (e.g. Powerpoint, Keynote, etc.), a data projector, a document camera, or a wireless hub that can project students' screens for the class. Communication through technology also includes the ability to share data sets or other mathematical results and to work collaboratively using internet chatrooms, virtual environments (see for example Muir, 2012) or other social networking programs. Clark-Wilson (2010) investigated the potentiality of network-technology using a wireless hub to connect calculators. Collection of classroom data through screen capture or running a quick poll of individual student answers, with the technology providing an aggregate of the class results, was found to support meaningful mathematical classroom discourse.

Communication *of* technology refers to the communication fostered by the mathematics generated by the technology (i.e. stimulus for discussion of the mathematics via outputs "of the technology" and of the mathematics behind a technological output). A key aspect here is that the technological output (such as that generated with CAS, dynamic geometry, applets or statistical tools) is prompting communication about mathematical concepts, skills, relationships, etc. The technology supports quick generation of mathematical objects (e.g. function graphs, geometric figures, algebraic outputs, statistical summaries, diagrams and mathematical applets) to promote mathematical discussion and investigation.

Schneider (2000) in a study comparing CAS classrooms and non-CAS classrooms in Year 11 described an evolving classroom dialogue in the CAS-classrooms. She found that the balance in who was leading communication in the classroom was different in the two types of classrooms. The shift in the CAS classroom was to less teacher-led communication ($\sim 80$ to $\sim 54$ %), with a greater focus on communication led by students, changing the dynamics in the classroom. Yackel (2002) presented results from a number of studies, in a range of contexts, related to the role of a teacher in collective argumentation in a class. She noted the important role of teachers in "initiating the negotiation of classroom norms that foster argumentation as the core of students' mathematical activity" (p. 423). In the context of classrooms with technology, teachers can consider new socio-mathematical norms to promote argumentation. Neill and Maguire (2006) reported a study of twelve Year 9 classes in a New Zealand CAS pilot where the classrooms included exploratory work focusing on understanding, rather than on rules and incorporated a range of teaching approaches including class discussion. Woo and Reeves (2006) suggest that although interaction is an essential component for learning, not every interaction results in learning. Therefore, it is important to understand the nature of interaction in the presence of technology and to investigate practices that promote student learning. Woo and Reeves suggest that negotiating, providing arguments, and considering various perspectives are crucial interactions. This is not a new consideration that arises as a result of technology, as there has been extensive research in the past highlighting the importance of argumentation in developing mathematical meaning (see for example, Wood, 1999). In trying to understand the role of technology in interactions during the process of learning and teaching, the new consideration is to conceptualize how teachers can take advantages of the affordances of technology to promote deep learning by students (see for example, Brown, 2010). Deep learning occurs in a classroom where teaching activities and student tasks promote speaking about mathematics, questioning results, applying known mathematics to unfamiliar situations, and reasoning in the exploration and solving of challenging mathematical ideas and problems.

Tasks that enable exploration of more examples, as well as consideration of more complex examples in the presence of technology, can provide material to discuss in class that foster development of meanings and models of concepts and promote deep learning.

The ability to use technology to produce results, to provide immediate feedback and to provide novel ways of looking at mathematical objects (e.g., consideration of applets for Pythagoras' theorem that show a visual proof; investigation of 3D objects in a virtual world; and consideration of dynamic visualizations of data) could support a change in the nature of communication in mathematics. Because students and teachers may be constrained by the availability of physical objects and by the nature of current resources, without technology, they can explore only a limited range of mathematical objects. With technologies, the range of mathematical objects and representations available is wider and more diverse; there are many more examples, models, and virtual experiences available to highlight different features of a mathematical relationship or concept. These examples, generated by technology and often preceded by experiences with physical objects, can lead to conjectures when students and teachers explore invariance and variance, look for patterns, and try to reason 'why' a particular result has occurred (Vincent, 2003). Consideration of mathematical ideas from a range of perspectives provides the potential to deepen students' understanding of mathematics when classroom discussions focus on making sense of the mathematical outputs from technology. In a study of Year 5/6 students using Tinkerplots, Fitzallen and Watson (2010) found that the students who had no experience with data handling were able to use Tinkerplots to support the development of statistical reasoning, with a range of representations used by different students. In a study investigating K-8 teachers' use of virtual manipulatives in classroom, Moyer-Packenham, Salkind and Bolyard (2008) found that teachers used virtual manipulatives to help students to learn mathematics content, focusing on investigation and skill consolidation. Leong and Koh (2003) found that when students in their study were learning using an inquiry approach, "richness in the development of the mathematical language and geometric understanding within these students in the process of their learning" resulted (p. 46). Here, an increase in communication was found to enhance mathematical communication and understanding. The teacher plays an important role in moderating and leading classroom discussion by identifying necessary prompts, providing promising inputs, or taking advantage of "hiccups" (Clark-Wilson, 2010). Sinclair and Robutti (2013) reported the essential role of the teacher in two case studies in which the teacher helped students to make links between dynamic representations and theoretical geometry.

To prepare teachers for the complexity of negotiating meaningful communication in the classroom it is essential to focus on associated issues and challenges in teacher professional development.

## 2.4 Communication and Collaboration: Professional Development

Over the past decade there have been many studies on the efficacy of professional development in education. Several literature reviews and meta-analyses summarize evidence for the efficacy of various projects in professional development (Lipowsky, 2010; Lipowsky & Rzejak, 2012; Timperly, Wilson, Barrar, & Fung, 2007; Yoon, Duncan, Lee, Scarloss, & Shapley, 2007). These publications identify different design principles for effective professional development. Professional communities that motivate reflection about teaching scenarios, student thinking and teacher beliefs as well as promote collaborative design of tasks and lessons are important for improving mathematics teaching (Llinares & Krainer, 2006; Rösken-Winter et al., 2015).

Neill and Maguire (2006) found that a positive outcome of the teacher professional development for the New Zealand CAS pilot was the collaboration between teachers. Although the teachers found that adopting a more exploratory based pedagogy and learning the technology was a time consuming aspect of the project, the "teachers and students reported changes to a more student-led, interactive, exploratory, collaborative, discussion-based style of teaching and learning" (p. ix).

Similar to the distinction between the two types of communication for teaching and learning, there are also two aspects of communication for professional development: communication *through* technology to foster collaboration (between teachers, as well as teachers and mathematics education researchers) and communication *of* technology as the focus of professional development sessions. Communication *through* technology can foster informal collaborations and support formal collaboration such as the type of collaboration that occurs in professional development (Kleickmann et al., 2012). Technologies to enable real-time-exchange (e.g., Skype, email) and collaborative work in shared spaces (like Moodle) can enhance informal collaborations in a school setting, as teachers interact with colleagues to reflect on their teaching experiences, share resources, and discuss teaching strategies. Access to online media for communication and the imminent viability of alternative forms of delivery, such as virtual learning environments (VLE), enable professional development to be more accessible to teachers to overcome issues of geographical distance (see for example, Ball, Steinle, & Chang, 2015). This is crucial for teachers who work in remote and rural locations where there may be a small number of mathematics teachers, or where face-to-face professional development is impossible due to distance.

Ball, Steinle, and Chang (2015) reported the trial of a prototype VLE for mathematics teacher professional development where the scripted interactions between an avatar teacher and an avatar student provided insight into the student's mathematical thinking. Participants in the trial commented that a VLE could be utilized in mathematics staff meetings to promote discussion amongst teachers about students' mathematical thinking. As new technologies, such as virtual learning environments, become available for teaching mathematics and even for delivering professional development there will be new opportunities to consider ways to communicate with teachers and also to promote communication among teachers to improve mathematics teaching on a broad scale.

Koellner, Jacobs, and Borko (2011) highlighted the importance of teachers reflecting on their classroom practice as a key component of professional growth. Teachers' use of video to collect data, to provide case studies to highlight effective pedagogies or to reflect on their own classes will be useful for promoting such reflection. Clarke (2006) described an approach for collection of rich classroom data as part of the Learners' Perspective Study, in which videos of a teacher's classroom prompted reflection about teaching and learning. Ho, Leong, and Ho (2015) investigated whether pre-service teachers' analysis of video of classroom use of an innovation could result in uptake of the innovation. They found that video provided a way for participants to see that a teaching innovation was not too time consuming and that it could be achieved in a normal classroom, with observable learning outcomes for students. Use of video case studies and purposeful self-reflection had the goal of promoting deeper reflection about teaching through observation of a classroom and reflection on personal beliefs about teaching.

Communication *of* technology is an important consideration in professional development sessions to prepare teachers for integration of technology into the classroom. One goal of professional development is to motivate an improvement in the pedagogical content knowledge of teachers (in this case, in using technology for teaching) with a subsequent positive impact on students' learning outcomes. In two studies in Singapore it was found that the use of Sketchpad helped students to identify "underlying geometric relationships in ways which are difficult to achieve with conventional static diagrams" (Ng & Leong, 2009, p. 306). Such examples of insights into learning with technology can provide motivation for teachers to consider new approaches for teaching mathematics with technology. These new dynamic opportunities that contrast with traditional (often static) approaches should be capitalized on in professional development sessions.

# 3   Summary and Looking Ahead

In previous sections of this survey the potential impact of technology on curricula, teaching, and learning in lower secondary mathematics has been highlighted. The mere presence of technology, however, does not guarantee improved teaching or learning. The challenge for teachers and students, therefore, is to utilize technology to improve learning outcomes in mathematics. In a technology-rich classroom, the teacher will play a pivotal role in crafting effective lessons that capitalize on the affordances of technology (Yerushalmy & Botzer, 2011). A key to planning and delivering effective lessons is to have good pedagogical content knowledge, which includes deep knowledge of students' understanding and how technology can positively influence this.

Lessons must include challenging and thought-provoking activities and tasks, and teachers should pay careful attention to classroom organization in order to foster mathematical thinking (Barzel 2006; Drijvers, Doorman, Boon, Reed, & Gravemeijer, 2010; Guin & Trouche, 1999; Trouche, 2004). Classroom discussions in which technology outputs act as catalysts for development of understanding can foster student learning (Pierce, Stacey, Wander, and Ball, 2011). One great benefit of using technology is that it can challenge teaching practices that promote mathematics only as a subject solely focusing on procedures, and foster mathematics as a subject incorporating conceptual understanding. For example, the use of CAS may promote students' consideration of the meaning of equivalence (Kieran & Drijvers, 2006). Rich mathematical experiences for students include ones that use dynamic representations (Sinclair & Yurita, 2008) and incorporate mathematical modelling (William & Goos, 2013). Technology provides a range of such possibilities. Current technology has provided opportunities for teachers to more easily obtain fine-grained and robust (i.e. based on research and experience) diagnostic assessment of student' conceptual understanding (Steinle & Stacey, 2012). In addition, the ability to report class-based results enables more targeting of lessons to address individual student needs (smartvic.com is one example of an environment that offers such features).

Over the past decade, the growth of technological resources for mathematics education has been rapid, with the development of software and hardware that has provided opportunities to change the nature of teaching and learning in lower secondary school mathematics. Predicting the role of technology in mathematics education in 2025 is a difficult task as we can only begin to imagine future technologies and the ways that current technologies will evolve to provide even greater possibilities for learning and teaching. The difficulty in predicting future technologies makes it impossible to predict what might be possible in mathematics education, but there is an opportunity to conceptualize principles that could provide directions for future technological developments.

We envisage that technology will be readily available for classroom use through affordable personal devices or even in the virtual world. Following is one vision for the potential role of technology in lower secondary mathematics in 2025.

We anticipate that future technologies will provide further opportunities to enrich mathematics education, in face-to-face or blended form (Clark-Wilson & Hoyles, 2015), at the lower secondary level in the following ways.

- Access to technology via personal devices will increase, with the consequence that technology integration into mathematics education, within and outside the classroom, can be easily realized. Students will have personal technology (e.g. a tablet, a smart watch, a mobile phone or similar) with which they are familiar, and, as a consequence, there will be less overhead in learning to use mathematics focused applications.
- The process of inputting mathematical entities will be intuitive and simple; mathematics on the screen can be easily manipulated, rather than being static. We hope that entering a complex equation into a technology will be as simple as writing it with pen and paper.
- Use of virtual experiences can enrich the exploration of virtual mathematical objects and concepts. Students will use technology to create mathematical objects, such as 3D objects, as well as use technology for analyzing those objects.
- Access to large data sets of interest to students will be more easily available. New ways to represent statistical data where students are part of a virtual presentation provide a rich experience of the nature of data. Virtual environments and augmented reality may provide the potential for students to be immersed in statistics, when they investigate the effect of changing foci, the impact of sampling and how different mathematical models can affect outcomes. For example, a model for predicting weather based on collected data might be explored. This perspective can facilitate a more experimental entrance to mathematics.
- The Internet of Things (IoT) is a term used to describe the next step in the evolution of smart objects—interconnected things in which the line between the physical object and digital information about that object is blurred. This stimulates improved opportunities for mathematical modelling and connections with the real world, through the ability to explore real situations in a virtual world. One implication of the IoT for lower secondary mathematics is the possibility to explore digital representations of concrete materials and to make connections to manipulation of concrete materials, in order to develop conceptual understanding. We imagine models that are either not possible or too expensive in real life becoming increasingly accessible. Integrated technologies will make exploration more accessible. For example, students could have access to technologies that allow them to explore different aspects of a 3D object by breaking it into parts, slicing through the object, and opening a net or rotating an object in space.
- E-textbooks, informed by mathematics education research, can fully integrate technologies designed to develop conceptual understanding: technologies that provide adaptive tasks that are tailored to an individual student based on their responses; diagnostic assessment that is embedded within technology; and technology that is used to provide research-informed diagnoses of thinking to a

teacher or a student in real time. Future mathematics classrooms could be paperless with students having access to devices on which they can easily document their mathematical work. For a true paperless classroom to be achieved, it is crucial for written communication of mathematics to be simple on a technological device, with technological devices that have the capacity to interpret handwriting, for example, during a diagnostic test.

- The ability for real time communication of mathematical thinking through synchronous communication technologies and virtual learning environments enable collaboration with virtual teachers, students, and others (such as scientists, researchers, etc.) regardless of location.

Where technology is readily available there is an imperative to provide professional development to deepen teachers' pedagogical content knowledge for teaching with technology. In 2025 we envisage strong links between mathematics education researchers and teachers, with research-based professional development and teaching resources easily accessible. Professional development must be possible in flexible modes such as virtual learning environments (VLE) or online communities through communication technologies in order to positively impact the learning of more students.

We assume that technology will be relatively simple for a teacher or a student to adapt for a specific mathematical purpose. The ability to shape and adapt learning objects to meet desired learning outcomes is often the realm of technical enthusiasts or programmers; however, with technological advances we anticipate that teachers will be able to customize technologies to meet their classroom needs.

We expect that in the next ten years there will be greater use of technology for delivery of effective professional development for teachers (e.g. through VLEs), as well as extensive professional development about teaching with technology. By 2025 we hope that synchronous and asynchronous modes of communication enable teachers to be part of a global community of mathematics educators, accessing information about best practice for mathematics teaching worldwide. New developments in the use of technology in mathematics education should be underpinned by research in mathematics education, and where appropriate, in combination with research into human computer interaction. The efficacy of innovations must be investigated, through valid and reliable research, to enable us to build on best practice.

Our overarching goal is that technology may be the vehicle to provide improved mathematics education for all students, regardless of location, school system, or other factors. The ability to extend the mathematical experience of students may be possible via a virtual world or exploration of online environments, or new software, or specific-purpose applets.

**Acknowledgement** The authors thank Alison Clark-Wilson, Brigitte Grugeon, Ulrich Kortenkamp, and Allen Leung for their highly appreciated input for this text.

# References

Adler, J. (2000). Conceptualising resources as a theme for teacher education. *Journal of Mathematics Teacher Education, 3*, 205–224.

Aldon, G., Cusi, A., Morselli, F., Panero, M., & Sabena, C. (in press). Formative assessment and technology: reflections developed through the collaboration between teachers and researchers In G. Aldon, F. Hitt, L. Bazzini, & U. Gellert (Eds.), *Mathematics and technology: A CIEAEM source book*. New York: Springer 'Advances in Mathematics Education'.

Anabousy, A., & Tabach, M. (2016). *Using GeoGebra to enhance students' inquiry activity*. Paper accepted for ICME13 TSG42.

Artigue, M. (2002). Learning mathematics in a CAS environment: The genesis of a reflection about instrumentation and the dialectics between technical and conceptual work. *International Journal of Computers for Mathematical Learning, 7*, 245–274.

Ball, L. (2015). *Use of computer algebra systems (CAS) and written solutions in a CAS allowed Year 12 mathematics subject: Teachers' beliefs and students' practices*. Unpublished doctoral dissertation, University of Melbourne, Melbourne, Australia.

Ball, L., Steinle, V., & Chang, S. (2015). A proof-of-concept virtual learning environment for professional learning of teachers of mathematics: Students' thinking about decimals. In K. Beswick, T. Muir, & J. Wells (Eds.), *Proceedings of the 39th Conference of the International Group for the Psychology of Mathematics Education* (Vol. 2, pp. 65–72). Hobart, Australia: PME.

Barzel, B. (2006). *Mathematik zwischen Konstruktion und Instruktion. Evaluation einer Lernwerkstatt 11 Jahrgang mit integriertem Einsatz Computeralgebra* (Dissertation). Essen: Universität Duisburg-Essen. http://duepublico.uni-duisburg-essen.de/servlets/DocumentServlet?id=13537

Barzel, B. (2012). *Computeralgebra – Mehrwert beim Lernen von Mathematik – aber wann?*. Münster: Waxmann Verlag.

Bell, B., & Cowie, B. (2001). The characteristics of formative assessment in science education. *Science Education, 85*(5), 533–536.

Bishop, J. P. (2013). Mathematical discourse as a process that mediates learning in SimCalc classrooms. In S.J. Hegedus & J. Roschelle (Eds.), *The SimCalc Vision and Contributions* (pp. 233–249). Berlin: Springer.

Black, P., & Wiliam, D. (1998). Assessment and classroom learning. *Assessment in education, 5*(1), 7–74.

Blume, G. W. & Heid, M. K. (Eds.) (2008). *Research on Technology and the Teaching and Learning of Mathematics: Vol. 2. Cases and Perspectives*. Charlotte, NC: Information Age.

Bokhove, C., & Drijvers, P. (2010). Digital tools for algebra education: Criteria and evaluation. *International Journal of Computers for Mathematical Learning, 15*(1), 45–62.

Brown, R. G. (2010). Does the introduction of the graphics calculator into system-wide examinations lead to change in the types of mathematical skills tested? *Educational Studies in Mathematics, 73*, 181–203.

Burrill, G., Allison, J., Breaux, G., Kastberg, S., Leatham, K., & Sanchez, W. (2002). *Handheld graphing technology at the secondary level: Research findings and implications for classroom practice*. Dallas, TX: Texas Instruments Corp. http://education.ti.com/research

Campuzano, L., Dynarski, M., Agodini, R., & Rall, K. (2009). *Effectiveness of reading and mathematics software products: Findings from two student cohorts—Executive summary* (NCEE 2009–4042). Washington, DC: National Center for Education Evaluation and Regional Assistance, Institute of Education Sciences, U.S. Department of Education.

Chenevotot-Quentin, F., Grugeon-Allys, B., Pilet J., & Delozanne, E. (accepted). Transfert du diagnostic Pépite à différents niveaux scolaires: Tests diagnostiques pour les élèves et leurs usages par les enseignant. *GT 6, Espace Mathématiques Francophone EMF 2015*, Alger, October 10–14, 2015.

Cheung, A. C. K., & Slavin, R. E. (2013). The effectiveness of educational technology applications for enhancing mathematics achievement in K-12 classrooms: A meta-analysis. *Educational Research Review, 9*, 88–113.

Clarke, D. J. (2006). The LPS Research Design. In D. J. Clarke, C. Keitel, & Y. Shimizu (Eds.), *Mathematics classrooms in twelve countries: The insider's perspective* (pp. 15–37). Rotterdam: Sense Publishers.

Clark-Wilson, A. (2010). *How does a multi-representational mathematical ICT tool mediate teachers' mathematical and pedagogical knowledge concerning variance and invariance* (Dissertation). London: Institute of Education, University of London.

Clark-Wilson, A., & Hoyles, C. (2015). Blended learning and e-learning support within the context of cornerstone maths—The changing culture of teachers' professional development. In K. Maaß, G. Törner, D. Wernisch, E. Schäfer, & K. Reitz-Koncebovski (Eds.), *Educating the educators: International approaches to scaling-up professional development in maths and science education* (pp. 158–166). Münster: Verlag für wissenschaftliche Texte und Medien.

Clements, M. A., Bishop, A., Keitel, C., Kilpatrick, J., & Leung, F. (Eds.). (2013). *Third international handbook of mathematics education*. New York: Springer.

Drijvers, P. (2004). Learning algebra in a computer algebra environment. *International Journal for Technology in Mathematics Education, 11*(3), 77–90.

Drijvers, P. (2009). Tools and tests: Technology in national final mathematics examinations. In C. Winslow (Ed.), *Nordic research on mathematics education, proceedings from NORMA08* (pp. 225–236). Rotterdam: Sense.

Drijvers, P. (2016). *Evidence for benefit? Reviewing empirical research on the use of digital tools in mathematics education*. Paper accepted for ICME13 TSG42.

Drijvers, P., Barzel, B., Maschietto, M., & Trouche, L. (2006). Tools and technologies in mathematical didactics. In M. Bosch (Ed.), *Proceedings of the Fourth Congress of the European Society for Research in Mathematics Education* (pp. 927–938). Barcelona, Spain: Universitat Ramon Llull. http://ermeweb.free.fr/CERME4/

Drijvers, P., Doorman, M., Boon, P., Reed, H., & Gravemeijer, K. (2010). The teacher and the tool: Instrumental orchestrations in the technology-rich mathematics classroom. *Educational Studies in Mathematics, 75*(2), 213–234.

Drijvers, P., & Trouche, L. (2008). From artifacts to instruments: A theoretical framework behind the orchestra metaphor. In G. W. Blume & M. K. Heid (Eds.), *Research on technology and the teaching and learning of mathematics* (Vol. 2, pp. 363–392)., Cases and perspectives Charlotte, NC: Information Age.

Dynarski, M., Agodini, R., Heaviside, S., Novak, T., Carey, N., & Campuzano, L. (2007). *Effectiveness of reading and mathematics software products: Findings from the first student cohort. Report to Congress*. Publication NCEE 2007-4005. Washington, DC: U.S. Department of Education.

EACEA/Eurydice. (2011). *Mathematics education in Europe: Common challenges and national policies*. Brussels: Eurydice.

Ellington, A. J. (2003). A meta-analysis of the effects of calculators on students' achievement and attitude levels in precollege mathematics classes. *Journal for Research in Mathematics Education, 34*(5), 433–463.

Ellington, A. J. (2006). The effects of non-CAS graphing calculators on student achievement and attitude levels in mathematics: A meta-analysis. *School Science and Mathematics, 106*(1), 16–26.

Even, R., & Ball, D. L. (Eds.). (2009). *The professional education and development of teachers of mathematics. The 15th ICMI Study. New ICMI Study Series,* (Vol. 11). New York/Berlin: Springer.

Falcade, R., Laborde, C., & Mariotti, M.-A. (2007). Approaching functions: Cabri tools as instruments of semiotic mediation. *Educational Studies in Mathematics, 66*(3), 317–333.

Fey, J. T., et al. (1984). *Computing and mathematics. The impact on secondary school curricula.* Reston, VA: National Council of Teachers of Mathematics.

Fitzallen, N., & Watson, J. (2010) Develoing statistical reasoning facilitated by Tinkerplots. In K. Makar, B. de Sousa, & R. Gould (Eds.), *Sustainability in statistics education. Proceedings of the Ninth International Conference on Teaching Statistics* (ICOTS9, July, 2014), Flagstaff, Arizona, USA. Voorburg, the Netherlands: International Statistical Institute. iase wcb.org

Fuglestad, A. B. (2009). ICT for Inquiry in mathematics: A developmental research approach. *Journal of Computers in Mathematics and Science Teaching, 28*(2), 191–202.

Graham, C. R. (2011). Theoretical considerations for understanding technological pedagogical content knowledge (TPACK). *Computers & Education, 57*, 1953–1960.

Grugeon, B. (2016). *Online automated assessment and student learning: The Pépite project in elementary algebra.* Paper accepted for ICME13 TSG42.

Grugeon, B., Chenevotot, F., Pilet, J., & Delozanne, E. (2013). Development and use of a diagnostic tool in elementary algebra using an online item bank. In B. Barton & S. Je Cho (Eds.), *Proceedings of International Congress on Mathematical Education, ICME 2012, Séoul, Corée, du 8 au 15 juillet 2012.*

Guin, D., & Trouche, L. (1999). The complex process of converting tools into mathematical instruments: The case of calculators. *International Journal of Computers for Mathematical Learning, 3*, 195–227.

Guo, K., & Cao, Y. (2012). A comparative study of the information technology use in Mathematics Curriculum in 14 countries. *China Educational Technology, 7*, 108–113.

Guo, K., & Cao, Y. (2015). Survey of mathematics teachers' technological pedagogical content knowledge and analysis of the influence factors. *Educational Science Research, 3*, 41–48.

Hadas, N., Hershkowitz, R., & Schwarz, B. B. (2000). The role of contradiction and uncertainty in promoting the need to prove in dynamic geometry environments. *Educational Studies in Mathematics, 44*(1–2), 127–150.

Heeren, B., & Jeuring, J. (2014). Feedback services for stepwise exercises. *Science of Computer Programming, 88*, 110–129.

Hegedus, S. J., & Roschelle, J. (Eds.). (2012). *The SimCalc vision and contributions: Democratizing access to important mathematics.* New York: Springer.

Heid, M. K. (1997). The technological revolution and the reform of school mathematics. *American Journal of Education, 106*(1), 5–61.

Heid, M. K., & Blume, G. W. (Eds.) (2008). *Research on technology and the teaching and learning of mathematics: Volume 1. Research syntheses.* Charlotte, NC: Information Age.

Higgins, S., Xiao, Z., & Katsipataki, M. (2012). *The impact of digital technology on learning: A summary for the education endowment foundation. Report.* Durham: Durham University.

Ho, W. K., Leong, Y. H., & Ho, F. H. (2015). The impact of online video suite on the Singapore pre-service teachers' buying-into innovative teaching of factorisation via AlgeCards. In S. W. Ng (Ed.), *Cases of mathematics professional development in East Asian countries: Using video to support grounded analysis* (pp. 157–178). Singapore: Springer.

Hoyles, C., & Lagrange, J.-B. (Eds.). (2010). *Mathematics education and technology–Rethinking the terrain.* New York / Berlin: Springer.

Hoyles, C., Noss, R., Kent, P., & Bakker, A. (2010). *Improving mathematics at work: The need for techno-mathematical literacies*. London: Routledge.

Kieran, C., & Drijvers, P. (2006). The co-emergence of machine techniques, paper-and-pencil techniques, and theoretical reflection: A study of CAS use in secondary school algebra. *International Journal of Computers for Mathematical Learning, 11*(2), 205–263.

Kleickmann, T., Richter, D., Kunter, M., Elsner, J., Besser, M., Krauss, S., & Baumert, J. (2012). Teachers' content knowledge and pedagogical content knowledge: The role of structural differences in teacher education. *Journal of Teacher Education, 64*(1), 90–106.

Koehler, M. J., Mishra, P., & Yahya, K. (2007). Tracing the development of teacher knowledge in a design seminar: Integrating content, pedagogy and technology. *Computers & Education, 49*, 740–762.

Koellner, K., Jacobs, J., & Borko, H. (2011). Mathematics professional development: Critical features for developing leadership skills and building teachers' capacity. *Mathematics Teacher Education and Development, 13*(1), 115–136.

Kulik, J. A. (2003). *Effects of using instructional technology in elementary and secondary schools: What controlled evaluation studies say*. Arlington, VA: SRI International. http://www.sri.com/policy/csted/reports/sandt/it/Kulik_ITinK-12_Main_Report.pdf

Lagrange, J.-B., Artigue, M., Laborde, C., & Trouche, L. (2003). Technology and mathematics education: a multidimensional study of the evolution of research and innovation. In A. J. Bishop, M. A. Clements, C. Keitel, J. Kilpatrick, & F. K. S. Leung (Eds.), *Second international handbook of mathematics education* (pp. 239–271). Dordrecht, Netherlands: Kluwer Academic Publishers.

Landry, S. H., Anthony, J. L., Swank, P. R., & Monseque-Bailey, P. (2009). Effectiveness of comprehensive professional development for teachers of at risk preschools. *Journal of Educational Psychology, 101*(2), 448–465.

Leong, Y. H., & Koh, L.-T. S. (2003). Effects of geometer's sketchpad on spatial ability and achievement in transformation geometry among two secondary students in Singapore. *The Mathematics Educator, 7*(1), 32–48.

Leung, A. (2015). Discernment and reasoning in dynamic geometry environments. In S. J. Cho (Ed.), *Selected regular lectures from the 12th international congress on mathematical education* (pp. 551–569). Switzerland: Springer.

Li, Q., & Ma, X. (2010). A meta-analysis of the effects of computer technology on school students' mathematics learning. *Educational Psychology Review, 22*, 215–243.

Lipowsky, F. (2010). Lernen im Beruf – Empirische Befunde zur Wirksamkeit von Lehrerfortbildung. In F. Müller, A. Eichenberger, M. Lüders, & J. Mayr (Eds.), *Lehrerinnen und Lehrer lernen – Konzepte und Befunde zur Lehrerfortbildung* (pp. 51–70). Münster: Waxmann.

Lipowsky, F., & Rzejak, D. (2012). Lehrerinnen und Lehrer als Lerner – Wann gelingt der Rollentausch? Merkmale und Wirkungen effektiver Lehrerfortbildungen. *Schulpädagogik heute, 5*(3), 1–17.

Llinares, S., & Krainer, K. (2006). Professional aspects of teaching mathematics. In A. Gutierrez & P. Boero (Eds.), *Handbook of research on the psychology of mathematics education. Past, present and future* (pp. 429–459). Rotterdam: Sense Publishers.

Luz, Y., & Yerushalmy, M. (2015). E-assessment of understanding of geometric proofs using interactive diagrams. In K. Krainer & N. Vondrová (Eds.), *Proceedings of CERME9* (pp. 149–155). Prague: Charles University.

Maschietto, M. (2016). *Classical and digital technologies for the Pythagorean theorem*. Paper accepted for ICME13 TSG42.

Maschietto, M., & Trouche, L. (2010). Mathematics learning and tools from theoretical, historical and practical points of view: the productive notion of mathematics laboratories. *ZDM-The International Journal on Mathematics Education, 42*(1), 33–47.

Moyer-Packenham, P. S., Salkind, G., & Bolyard, J. J. (2008). Virtual manipulatives used by K-8 teachers for mathematics instruction: Considering mathematical, cognitive, and pedagogical fidelity. *Contemporary Issues in Technology and Teacher Education, 8*(3), 202–218.

Muir, T. (2012). Virtual mathematics education: Using second life to model and reflect upon the teaching of mathematics. In J. Dindyal, L. P. Cheng, & S. F. Ng (Eds.), *Mathematics education: Expanding horizons. Proceedings of the 35th Annual Conference of the Mathematics Education Research Group of Australasia* (pp. 521–528). Singapore: MERGA.

Neill, A., & Maguire, T. (2006). *An evaluation of the CAS pilot project: Report for the ministry of education and the new zealand qualifications authority.* Wellington, NZ: New Zealand Council for Educational Research.

Ng, W. L., & Leong, Y. H. (2009). Use of ICT in mathematics education in Singapore: Review of research. In W. K. Yoong, L. P. Yee, B. Kaur, F. P. Yee, & N. S. Fong (Eds.), *Mathematics education: The Singapore journey.* Singapore: World Scientific.

OECD (2015). Students, computers and learning. making the connection. Paris: OECD Publishing. http://www.oecd.org/edu/students-computers-and-learning-9789264239555-en.htm

Pierce, R., & Stacey, K. (2010). Mapping pedagogical opportunities provided by mathematics analysis software. *International Journal of Computers for Mathematical Learning, 15*(1), 1–20.

Pierce, R., Stacey, K., Wander, R., & Ball, L. (2011). The design of lessons using mathematics analysis software to support multiple representations in secondary school mathematics. *Technology, Pedagogy and Education, 20*(1), 95–112.

Rakes, C. R., Valentine, J. C., McGatha, M. B., & Ronau, R. N. (2010). Methods of instructional improvement in Algebra: A systematic review and meta-analysis. *Review of Educational Research, 80*(3), 372–400.

Remillard, J. T. (2005). Examining key concepts of research on teachers' use of mathematics curricula. *Review of Educational Research, 75*(2), 211–246.

Roberts, D., Leung, A., & Lin, B. (2013). From the slate to the web: Technology in the mathematics curriculum. In A. Bishop, M. A. Clements, C. Keitel, J. Kilpatrick, & F. Leung (Eds.), *Third international handbook of mathematics education* (pp. 525–547). Berlin: Springer.

Roesken, B. (2011). Mathematics teacher professional development. In B. Roesken (Ed.), *Hidden dimensions in the professional development of mathematics teachers* (pp. 1–28). Rotterdam: Sense Publishers.

Ronau, R. N., Rakes, C. R., Bush, S. B., Driskell, S. O., Niess, M. L., & Pugalee, D. K. (2014). A survey of mathematics education technology dissertation scope and quality: 1968–2009. *American Educational Research Journal, 51*(5), 974–1006.

Rösken-Winter, B., Schüler, S., Stahnke, R., & Blömeke, S. (2015). Effective CPD on a large scale: Examining the development of multipliers. *ZDM, 47*(1), 13–25.

Ruchniewicz, H. (2016). *Developing a digital tool for formative self-assessment.* Paper accepted for ICME13 TSG42.

Ruthven, K. (2007). Teachers, technologies and the structures of schooling. In D. Pitta-Pantazi & G. Philippou (Eds.), *Proceedings of the V congress of the european society for research in mathematics education CERME5* (pp. 52–67). Larnaca, Cyprus: University of Cyprus.

Ruthven, K. (2009). Towards a naturalistic conceptualisation of technology integration in classroom practice: The example of school mathematics. *Education & Didactique, 3*(1), 131–159.

Sacristán, A. I., Calder, N., Rojano, T., Santos, M., Friedlander, A., & Meissner, H. (2010). The influence and shaping of digital technologies on the learning—and learning trajectories—of mathematical concepts. In C. Hoyles & J.-B. Lagrange (Eds.), *Mathematics education and technology—Rethinking the terrain* (pp. 179–226). Dordrecht, The Netherlands: Springer.

Schneider, E. (2000). Teacher experiences with the use of CAS in a mathematics classroom. *The International Journal of Computer Algebra in Mathematics Education, 7*(2), 119–141.

Sinclair, N., & Robutti, O. (2013). Technology and the role of proof: The case of dynamic geometry. In A. Bishop, M. A. Clements, C. Keitel, J. Kilpatrick, & F. Leung (Eds.), *Third international handbook of mathematics education* (pp. 571–596). New York: Springer.

Sinclair, N., & Yurita, V. (2008). To be or to become: How dynamic geometry changes discourse. *Research in Mathematics Education, 10*(2), 135–150.

Stacey, K., & Wiliam, D. (2013). Technology and assessment in mathematics. In M. A. Clements, A. Bishop, C. Keitel, J. Kilpatrick, & F. Leung (Eds.), *Third international handbook of mathematics education* (pp. 721–751). New York: Springer.

Steinle, V., & Stacey, K. (2012). Teachers' views of using an on-line, formative assessment system for mathematics. *Pre-proceedings. 12th International Congress on Mathematical Education Topic Study Group 33* (pp 6721–6730). COEX, Seoul, Korea.

Timperley, H., Wilson, A., Barrar, H., & Fung, I. (2007). *Teacher professional learning and development. Best Evidence Synthesis Iteration.* Wellington, New Zealand: Ministry of Education.

Tokpah, C. L. (2008). The effects of computer algebra systems on students' achievement in mathematics (Doctoral dissertation). Kent, OH: Kent State University.

Trouche, L. (2004). Managing complexity of human/machine interactions in computerized learning environments: Guiding students' command process through instrumental orchestrations. *International Journal of Computers for Mathematical Learning, 9*, 281–307.

VanLehn, K. (2011). The relative effectiveness of human tutoring, intelligent tutoring systems, and other tutoring systems. *Educational Psychologist, 46*(4), 197–221.

Vincent, J. (2003). Year 8 students' reasoning in a Cabri environment. In L. Bragg, C. Campbell, G. Herbert, & J. Mousley (Eds.) *Mathematics education research: Innovation, networking, opportunity. Proceedings of the 26th Annual Conference of the Mathematics Education Research Group of Australasia, Geelong* (pp. 696–703). Sydney: MERGA.

Wiliam, D. (2011). What is assessment for learning? *Studies in Educational Evaluation, 37*(1), 2–14.

Wiliam, D., & Thompson, M. (2008). Integrating assessment with learning: What will it take to make it work? In C. A. Dwyer (Ed.), *The future of assessment: Shaping teaching and learning* (pp. 53–82). Mahwah, NJ: Lawrence Erlbaum Associates.

William, J., & Goos, M. (2013). Modelling with mathematics and technology. In A. Bishop, M. A. Clements, C. Keitel, J. Kilpatrick, & F. Leung (Eds.), *Third international handbook of mathematics education* (pp. 549–569). Berlin: Springer.

Woo, Y., & Reeves, T. C. (2006). *Meaningful online learning: Exploring interaction in a web-based learning environment using authentic tasks.* Unpublished Dissertation. University of Georgia, Athens, GA.

Wood, T. (1999). Creating a context for argument in mathematics class. *Journal for Research in Mathematics Education, 30*(2), 171–191.

Wools, S., Eggen, T., & Sanders, P. (2010). Evaluation of validity and validation by means of the argument-based approach. *CADMO, 8*, 63–82.

Yackel, E. (2002). What we can learn from analyzing the teacher's role in collective argumentation? *Journal of Mathematical Behavior, 21*, 423–440.

Yerushalmy, M., & Botzer, G. (2011). Teaching secondary mathematics in the mobile age. In O. Zaslavsky & P. Sullivan (Eds.), *Constructing knowledge for teaching secondary mathematics tasks to enhance prospective and practicing teacher learning* (pp. 191–208). New York: Springer.

Yoon, K. S., Duncan, T., Lee, S. W.-Y., Scarloss, B., & Shapley, K. L. (2007). *Reviewing the evidence on how teacher professional development affects student achievement.* Washington, DC: U.S. Department of Education.

Zbiek, R. M., Heid, M. K., Blume, G. W., & Dick, T. P. (2007). Research on technology in mathematics education: The perspective of constructs. In F. K. Lester Jr (Ed.), *Handbook of research on mathematics teaching and learning* (2nd ed., pp. 1169–1207). Charlotte, NC: Information Age Publishing.

Zehavi, N., & Mann, G. (1999). The expressive power provided by a solving tool: How long did Diophantus live? *International Journal of Computer Algebra in Mathematics Education, 6*(4), 249–266.

# Further Reading

Blume, G. W., & Heid, M. K. (Eds.). (2008). *Research on technology and the teaching and learning of mathematics: Vol. 2. Cases and perspectives.* Charlotte, NC: Information Age.

Clements, M. A. Bishop, A., Keitel, C., Kilpatrick, J., & Leung, F. (Eds.). (2013). *Third international handbook of mathematics education.* New York: Springer.

Even, R., & Ball, D. L. (Eds.). (2009). *The professional education and development of teachers of mathematics. The 15th ICMI Study. New ICMI Study Series, Vol. 11.* New York/Berlin: Springer.

Heid, M. K., & Blume, G. W. (Eds.). (2008). *Research on Technology and the Teaching and Learning of Mathematics: Volume 1. Research Syntheses.* Charlotte, NC: Information Age.

Hoyles, C., & Lagrange, J-B. (Eds.). (2010). *Mathematics education and technology—Rethinking the terrain.* New York/Berlin: Springer.

OECD. (2015). *Students, computers and learning. Making the connection.* Paris: OECD Publishing. http://www.oecd.org/edu/students-computers-and-learning-9789264239555-en.htm

Pierce, R., & Stacey, K. (2004). Monitoring progress in algebra in a CAS active context: Symbol sense, algebraic insight and algebraic expectation. *International Journal for Technology in Mathematics Education, 11*(1), 3–11.

# Further Reading

Blume, G. W., & Heid, M. K. (Eds.). (2008). *Research on technology and the teaching and learning of mathematics: Vol. 2. Cases and perspectives*. Charlotte, NC: Information Age.

Clements, M. A. Bishop, A., Keitel, C., Kilpatrick, J., & Leung, F. (Eds.). (2013). *Third international handbook of mathematics education*. New York: Springer.

Even, R., & Ball, D. L. (Eds.). (2009). *The professional education and development of teachers of mathematics. The 15th ICMI Study. New ICMI Study Series, Vol. 11*. New York/Berlin: Springer.

Heid, M. K., & Blume, G. W. (Eds.). (2008). *Research on Technology and the Teaching and Learning of Mathematics: Volume 1. Research Syntheses*. Charlotte, NC: Information Age.

Hoyles, C., & Lagrange, J-B. (Eds.). (2010). *Mathematics education and technology—Rethinking the terrain*. New York/Berlin: Springer.

OECD. (2015). *Students, computers and learning. Making the connection*. Paris: OECD Publishing. http://www.oecd.org/edu/students-computers-and-learning-9789264239555-en.htm

Pierce, R., & Stacey, K. (2004). Monitoring progress in algebra in a CAS active context: Symbol sense, algebraic insight and algebraic expectation. *International Journal for Technology in Mathematics Education, 11*(1), 3–11.

Stacey, K., & Wiliam, D. (2013). Technology and assessment in mathematics. In M. A. Clements, A. Bishop, C. Keitel, J. Kilpatrick, & F. Leung (Eds.), *Third international handbook of mathematics education* (pp. 721–751). New York: Springer.

Steinle, V., & Stacey, K. (2012). Teachers' views of using an on-line, formative assessment system for mathematics. *Pre-proceedings. 12th International Congress on Mathematical Education Topic Study Group 33* (pp 6721–6730). COEX, Seoul, Korea.

Timperley, H., Wilson, A., Barrar, H., & Fung, I. (2007). *Teacher professional learning and development. Best Evidence Synthesis Iteration.* Wellington, New Zealand: Ministry of Education.

Tokpah, C. L. (2008). The effects of computer algebra systems on students' achievement in mathematics (Doctoral dissertation). Kent, OH: Kent State University.

Trouche, L. (2004). Managing complexity of human/machine interactions in computerized learning environments: Guiding students' command process through instrumental orchestrations. *International Journal of Computers for Mathematical Learning, 9,* 281–307.

VanLehn, K. (2011). The relative effectiveness of human tutoring, intelligent tutoring systems, and other tutoring systems. *Educational Psychologist, 46*(4), 197–221.

Vincent, J. (2003). Year 8 students' reasoning in a Cabri environment. In L. Bragg, C. Campbell, G. Herbert, & J. Mousley (Eds.) *Mathematics education research: Innovation, networking, opportunity. Proceedings of the 26th Annual Conference of the Mathematics Education Research Group of Australasia, Geelong* (pp. 696–703). Sydney: MERGA.

Wiliam, D. (2011). What is assessment for learning? *Studies in Educational Evaluation, 37*(1), 2–14.

Wiliam, D., & Thompson, M. (2008). Integrating assessment with learning: What will it take to make it work? In C. A. Dwyer (Ed.), *The future of assessment: Shaping teaching and learning* (pp. 53–82). Mahwah, NJ: Lawrence Erlbaum Associates.

William, J., & Goos, M. (2013). Modelling with mathematics and technology. In A. Bishop, M. A. Clements, C. Keitel, J. Kilpatrick, & F. Leung (Eds.), *Third international handbook of mathematics education* (pp. 549–569). Berlin: Springer.

Woo, Y., & Reeves, T. C. (2006). *Meaningful online learning: Exploring interaction in a web-based learning environment using authentic tasks.* Unpublished Dissertation. University of Georgia, Athens, GA.

Wood, T. (1999). Creating a context for argument in mathematics class. *Journal for Research in Mathematics Education, 30*(2), 171–191.

Wools, S., Eggen, T., & Sanders, P. (2010). Evaluation of validity and validation by means of the argument-based approach. *CADMO, 8,* 63–82.

Yackel, E. (2002). What we can learn from analyzing the teacher's role in collective argumentation? *Journal of Mathematical Behavior, 21,* 423–440.

Yerushalmy, M., & Botzer, G. (2011). Teaching secondary mathematics in the mobile age. In O. Zaslavsky & P. Sullivan (Eds.), *Constructing knowledge for teaching secondary mathematics tasks to enhance prospective and practicing teacher learning* (pp. 191–208). New York: Springer.

Yoon, K. S., Duncan, T., Lee, S. W.-Y., Scarloss, B., & Shapley, K. L. (2007). *Reviewing the evidence on how teacher professional development affects student achievement.* Washington, DC: U.S. Department of Education.

Zbiek, R. M., Heid, M. K., Blume, G. W., & Dick, T. P. (2007). Research on technology in mathematics education: The perspective of constructs. In F. K. Lester Jr (Ed.), *Handbook of research on mathematics teaching and learning* (2nd ed., pp. 1169–1207). Charlotte, NC: Information Age Publishing.

Zehavi, N., & Mann, G. (1999). The expressive power provided by a solving tool: How long did Diophantus live? *International Journal of Computer Algebra in Mathematics Education, 6*(4), 249–266.